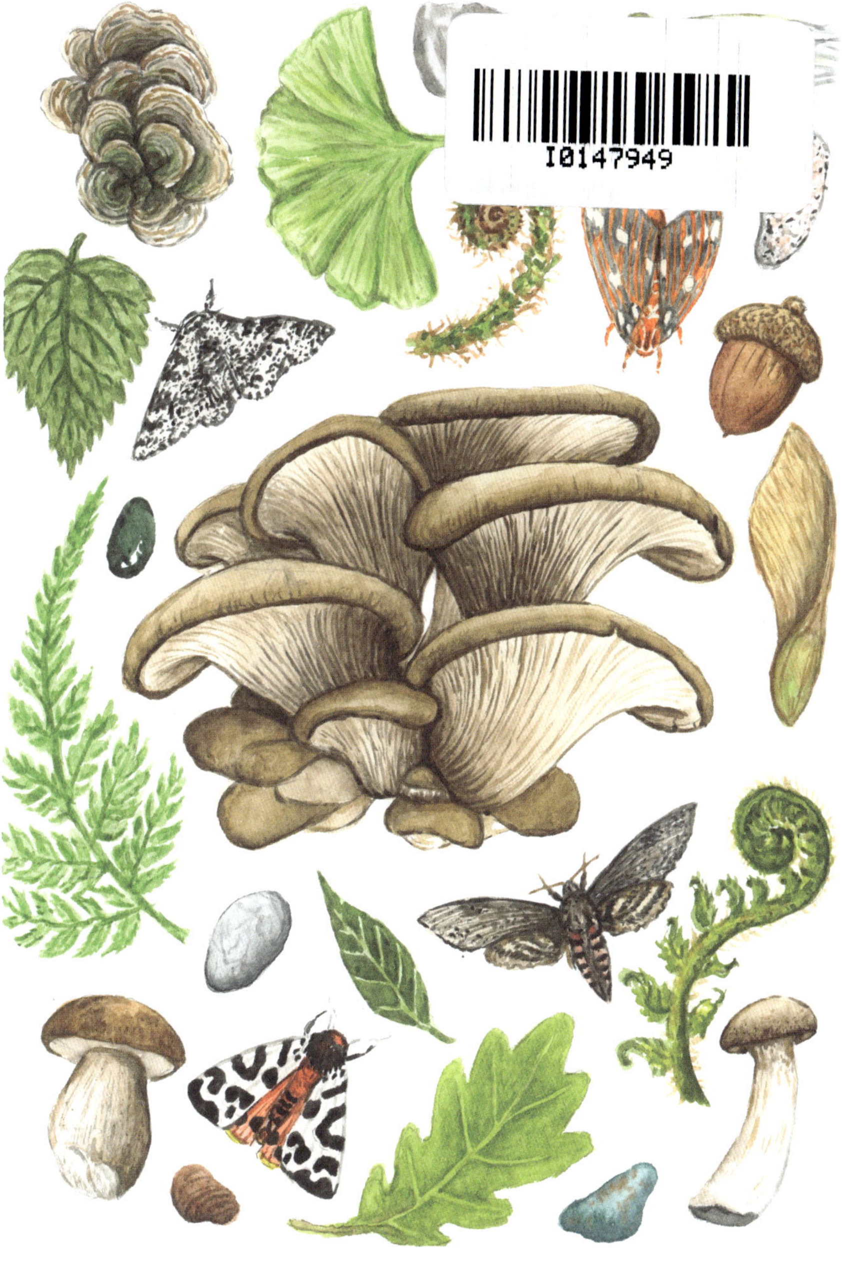

THE WONDER

of

MUSHROOMS

The Mysterious World of Fungi

MAYA JEWELL ZELLER

ILLUSTRATED BY JENNY DeFOUW GEUDER

Adventure PUBLICATIONS

SAFETY NOTE: *The world of fungi is fantastic. This book is a gentle introduction to the marvels and wonders within, but it is not a field guide, nor is it a guide to foraging fungi and mushrooms.*

Cover and text design: Hilary Harkness
Editors: Brett Ortler and Andrew Mollenkof
Proofreader: Emily Beaumont
Technical reviewer: Stephen Brockman, PhD

10 9 8 7 6 5 4 3 2

The Wonder of Mushrooms
Copyright © 2025 by Maya Jewell Zeller (text);
Jenny DeFouw Geuder (illustrations)

LCCN 2025022758 (print); 2025022759 (ebook)
ISBN 978-1-64755-466-8 (pbk.); 978-1-64755-467-5 (ebook)

Published by Adventure Publications
An imprint of AdventureKEEN
310 Garfield Street South
Cambridge, MN 55008
(800) 678-7006
adventurepublications.net

TABLE OF CONTENTS

2 A Kingdom of Fungi

20 A Rainbow of Color

38 A Myriad of Shapes

64 Forms & Caps & Gills

83 Crinkle Skirts to a Fan That Frills: Gills

86 And Some Have Pores . . .

92 Spore Prints as Upside-Down Flowers

103 A Web Beneath the Forest Floor

116 Mycelial Multitudes

128 The Warm and Necessary Breath of Decomposition

137 Worlds unto Themselves

144 A Fecund Celebration of Death

155 A Brief History of Foxfire

167 Treasure from the Forest

172 Of Mushrooms and Mordants

185 Witches Know What's What

194 Inside the Fairy Ring

206 An Unexpected Savior?

213 A Sampling of Words for "Mushroom" from Around the World

214 Mushroom Names

216 About the Author

217 About the Illustrator

220 Works Cited and Further Reading

ACKNOWLEDGMENTS

We offer so much gratitude to Brett Ortler for envisioning the mushroom fruit of this work and for shepherding this book along its mycelial path.

Maya Jewell Zeller is thankful to her son, Canyon, for sleuthing out, as part of his natural 11-year-old curiosity, the information about engineering and architecture's exploration of hyphae structure; to Ellie Kozlowski for sharing the fungal energies; and to the literary predecessors whose works inform her wordplay:

"The black trumpet's strange form startles enough to make one stop for them" is an allusion to Emily Dickinson's "Because I could not stop for death . . ." (Note: Dickinson also wrote, "The Mushroom is the Elf of Plants.")

In "Mycelial Multitudes," the line "These networks contain multitudes" is an allusion to Walt Whitman.

Sylvia Plath said that mushrooms would inherit the earth.

A KINGDOM OF
FUNGI

ANIMALIA, PLANTAE, FUNGI,
PROTISTA, ARCHAEA OR
ARCHAEBACTERIA, AND BACTERIA
OR EUBACTERIA.

This isn't a prayer to the divine,
        though say it aloud a few times in a row
and you may feel a bit like a poet or priest.

This beautiful litany of words forms the
Linnean system of six kingdoms—by which
we currently classify biological
life in North America.

THESE MAY SOUND LIKE
ENCHANTING COMMUNITIES,
REIGNED BY QUEENS AND WITCHES,

but the truth is, they are mostly ruled
(read "ruled" as "made possible") by the
fungi, an expansive kingdom best known by
mushrooms, which are fungal fruiting bodies.

FASCINATINGLY,
FUNGI ARE MORE CLOSELY
RELATED TO ANIMALS
THAN PLANTS—

and there are perhaps
**150,000 recorded living
species,** and millions likely
yet undiscovered!

In other words,
what we know of fungi
is dwarfed by what we don't.

Of the ones we know,

fewer than 10% (about 14,000)

produce fruiting bodies,

or what we think of as mushrooms.

Of those species that produce
mushrooms, there are two primary groups:
BASIDIOMYCETES
AND ASCOMYCETES,
most with microscopic spores.

The basidiomycetes
(mushrooms, shelf fungi or polypores,
rusts, and smuts—from the Germanic
for dirt) have club-shaped spores;

ascomycetes
(truffles, cups, morels, and yeasts)
have cup- or sac-shaped spores
called ascus, from the Greek, askós,
or wineskin.

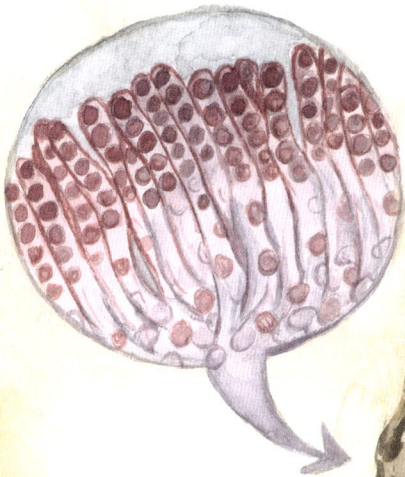

ASCOMYCETES

BASIDIOMYCETES

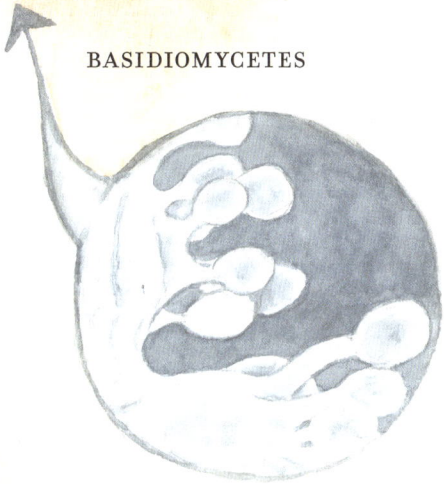

**FUNGI, AND PRODUCTS
MADE FROM THEM,**

are all around us, including
in beer, bread, and cheese.
Those are made possible
by **yeasts,** one of the fungi
groups that doesn't show up
as a fruiting body

(but which can certainly
be combined with fruits).

Other fungi help make
antibiotics like penicillin
and cephalosporins.

Consider the lichen:
a symbiosis between
algae or cyanobacteria
and—what else—
a fungus.

You've likely (lichen-ly?) seen lichen
and fungus growing on bark
or rocks, lining basalt canyons
with their orange-, yellow-, and
rust-colored crusts and dusts.

Those are part of the
fungi family tree too.

Art may err,

but Nature cannot miss.

—DRYDEN

A RAINBOW OF
COLOR

Like walls of samples at the
local paint store, the palette for fungi
fans out, hue after hue.
A library of mushrooms offers
a rainbow of colors:

TURKEY TAIL (sometimes called rainbow
fungus) is both an actual mushroom
    (which you might find on dead wood
        in your garden or in the forest)
and an inclusive concept—

   **you can witness a prism of color
   in the mycelium world,**

        ranging from the reds of
CRIMSON WAX CAP and the FLY AMANITA
    to the vivid yellow of DYER'S POLYPORE.

CRIMSON WAX CAP
    wears a red hat, with an orange
    glow of gills beneath its chin.

TURKEY
TAIL

CRIMSON
WAX CAP

FLY
AMANITA

DYER'S
POLYPORE

CORAL PINK MERULIUS

looks like a feverish
child's throat when
they say, "aah."

ORANGE MYCENA

comes in bright, its bells loud with orange,
then fades as it ages . . .
and only the edges of the gills
are actually orange; the insides are
cream-colored.

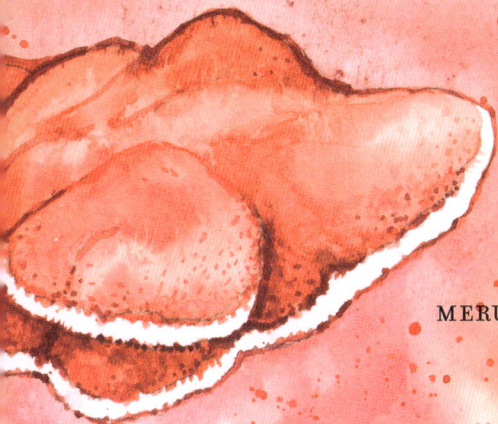

MERULIUS

MYCENA

25

CHICKEN OF THE WOODS

DYER'S POLYPORE

Chicken of the woods

BOASTS EDIBLE
    BRACKETS OF BRILLIANT
SUN-HUED GLOW.

Dyer's polypore
    begins yellow but ages
yellow-green or brown . . .

And those **DREAMY COLORS**
on a fallen log are a stain from

ELFCUP

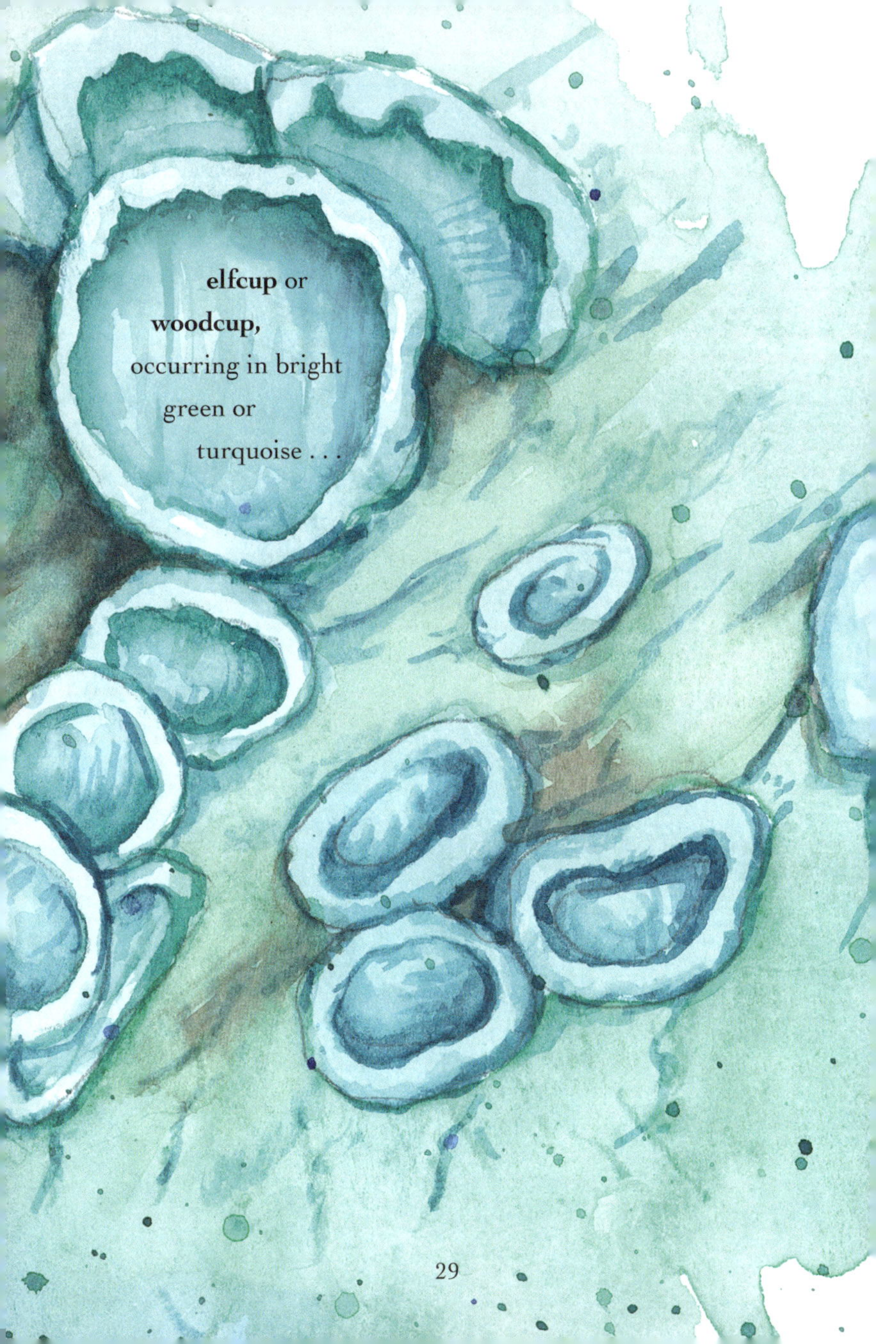

elfcup or
woodcup,
occurring in bright
green or
turquoise . . .

COBALT CRUST

On sunny sloping hills in June
erupt bachelor's buttons,
those beautiful flowers whose
color repeats almost exactly in
COBALT CRUST, also known as
VELVET BLUE SPREAD . . .

And sliding into
the royal
angle of the
rainbow, is
**VIOLET
WEBCAP,**
with its purple
stem, cap,
and gills.

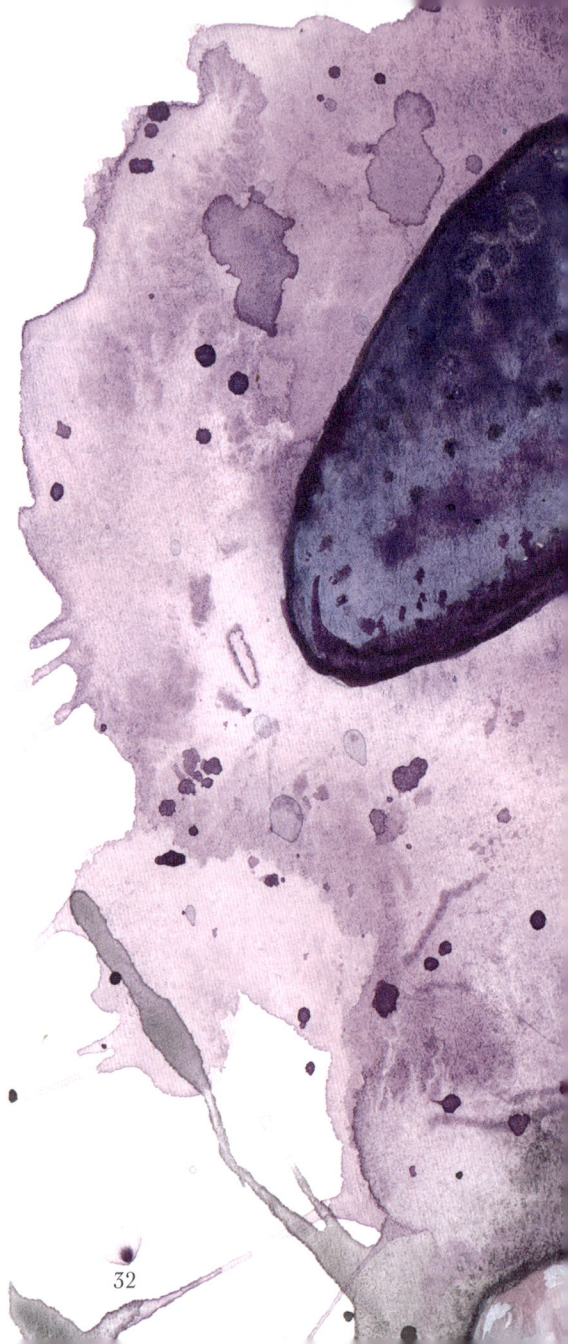

VIOLET
WEBCAP

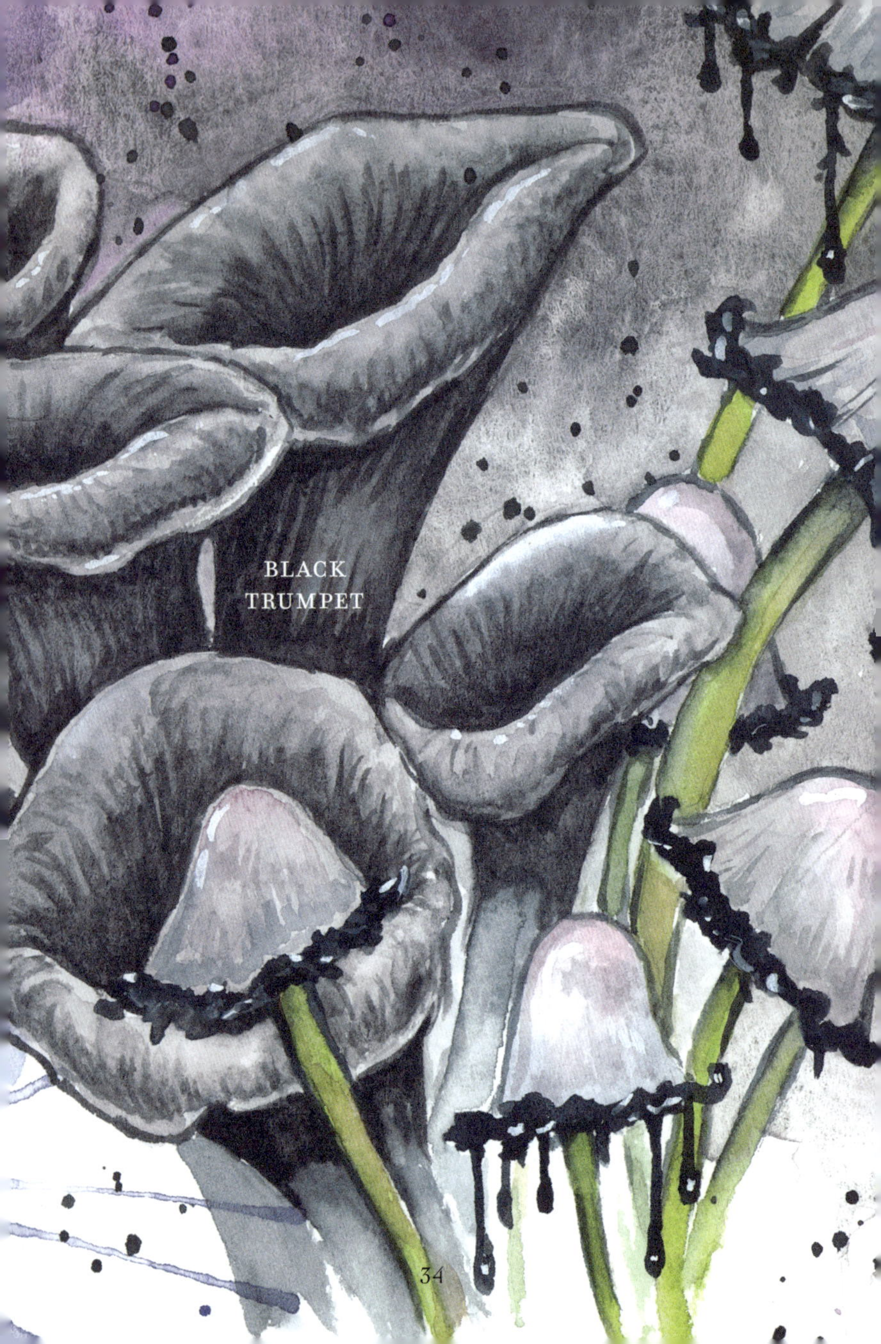
BLACK
TRUMPET

The
**BLACK
TRUMPET'S**
strange form
startles enough
to make one
stop for them,

INKY
CAPS

and
INKY CAPS
spring up
and
then
melt
away.

The golden-rod and asters adorn the roadsides, the odors of the sweet gale and scented fern are wafted gratefully to our senses as we pass along the lanes, and there, among the fallen leaves, at the very edge of the woods, peers out a bright yellow mushroom, brighter from the contrast to the dead leaves around, and then another, close by, and then a shining white cap; further on a mouse-colored one, gray, and silky in texture. What a contrast of colors. What are they? By what names shall we call them?

—Dallas and Burgin

A MYRIAD OF SHAPES

With all the colors of fungi, it's not surprising that they also come in myriad shapes. They range from the expected to the otherworldly, sometimes with names to match.

There are the classic-looking mushrooms
with a cap and stem, like a
GNOME IN A HAT.

There are also
SHELF MUSHROOMS,
which take another common shape—
and this group contains many
sourced for dye.

Shelf-shaped mushrooms
are usually polypores . . .

TURKEY
TAIL

TINDER POLYPORE

others in this same general

      shape contour include those

    that resemble **HOOVES,**

          such as the tinder polypore,

  or the **FAN-LIKE** turkey tail.

VEILED BOLETE

PUFFBALL

EARTHSTAR

SPHERICAL, round mushrooms, like puffballs,
bring a whimsy (and some even open
into stars, or contain a whole bolete
developing inside the veil).

CORAL-SHAPED

mushrooms resemble

something from under the sea,

as might CLUB or
DIGIT-SHAPED fungi like
dead man's fingers . . .

49

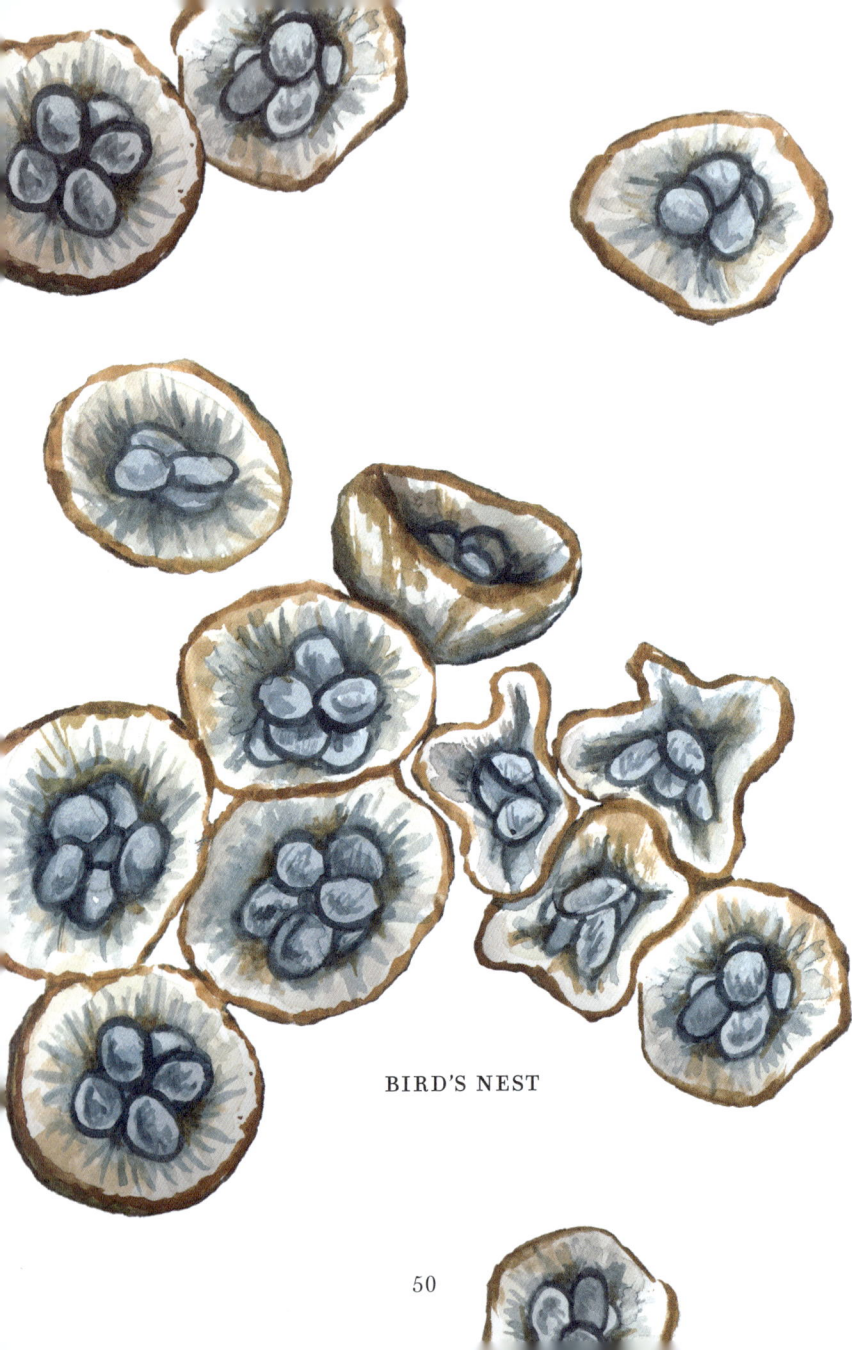

BIRD'S NEST

while bird's-nest fungi
and devil's urn look like
NESTS with eggs, or
CUPS, respectively.

DEVIL'S
URN

There are those that resemble **TULIPS**,
 such as the *Phallus* species, stinkhorns,
  and would fit well in the
Iceland museum of phallus shapes
 or inspire a myth or two.

STINKHORNS

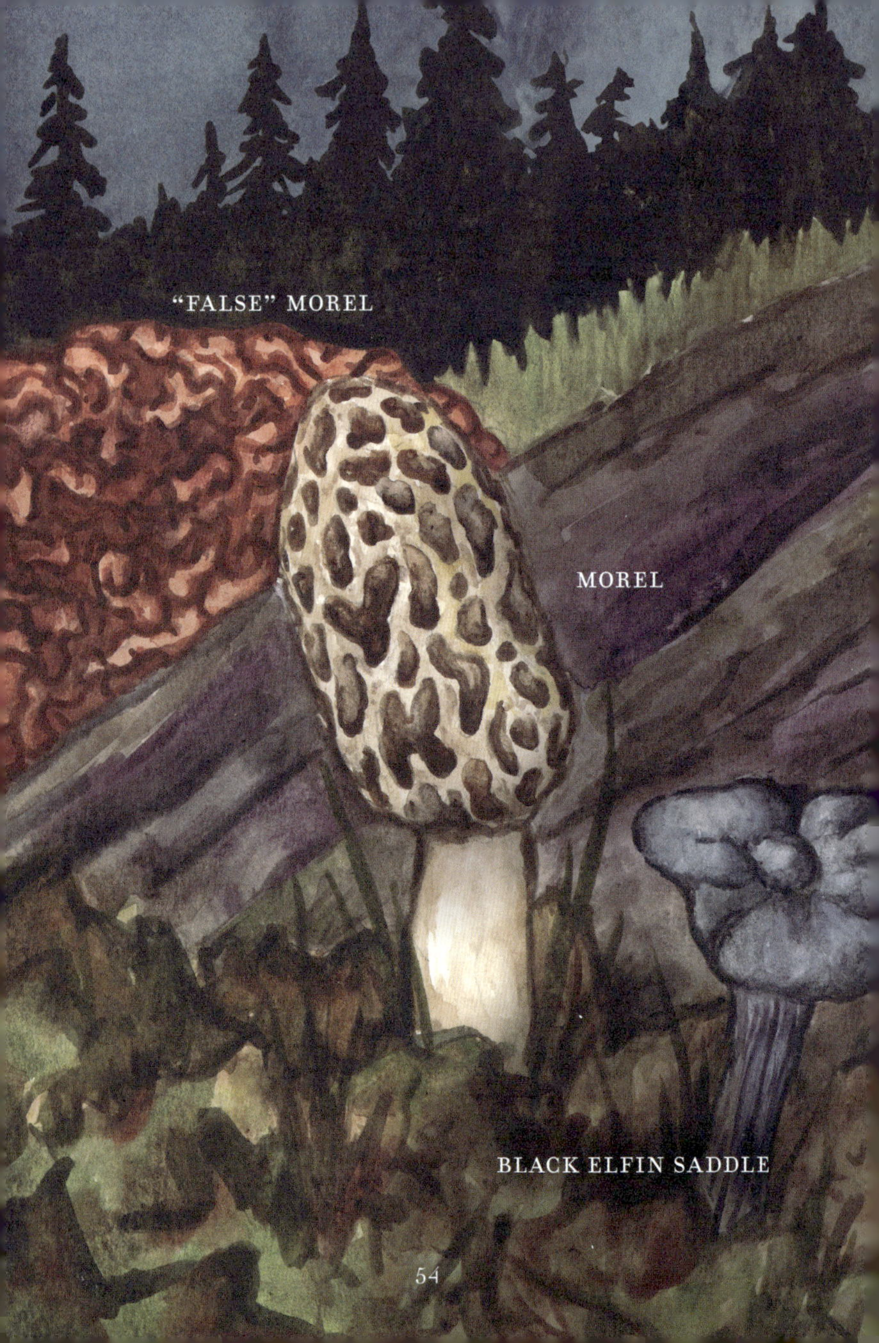
"FALSE" MOREL
MOREL
BLACK ELFIN SADDLE

And there are the **BRAIN-LIKE**
beauties: morels,
the misleadingly named "false" morels,
and the wonderfully named
Black Elfin Saddle.

And then those that imitate
MOLDS or JELLIES:
witch's butter, white jelly,
yellow fairy cups,
wrinkled peach, wood ear.

But the red cage stinkhorn
and column stinkhorn
are so ODD-SHAPED,
they look like something
sprung from a faraway planet.

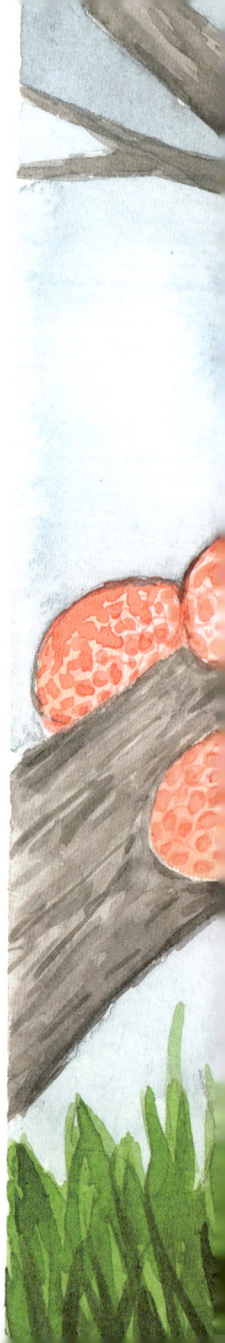

WITCH'S
BUTTER

WOOD EAR

WRINKLED
PEACH

COLUMN
STINKHORN

RED CAGE
STINKHORN

The **BLACK TRUMPET**

looks as if it might

honk out a squawk

or a screech—

AMANITA

60

and some mushrooms
  trade their looks over time.

Consider the Amanita,
  which starts out like an
  "EGG" and ends up
  looking quite different.

There is a wonderful enchantment
about these surprises of the young year.
For they are always surprises, never
mind how often we have experienced
them or how unfailingly we await them.

—ANNA BOTSFORD COMSTOCK

FORMS &
CAPS
& GILLS

When spotting a mushroom, consider
cataloging colors and
shapes and forms,
beginning with the details of a
mushroom's cap (if it has one).

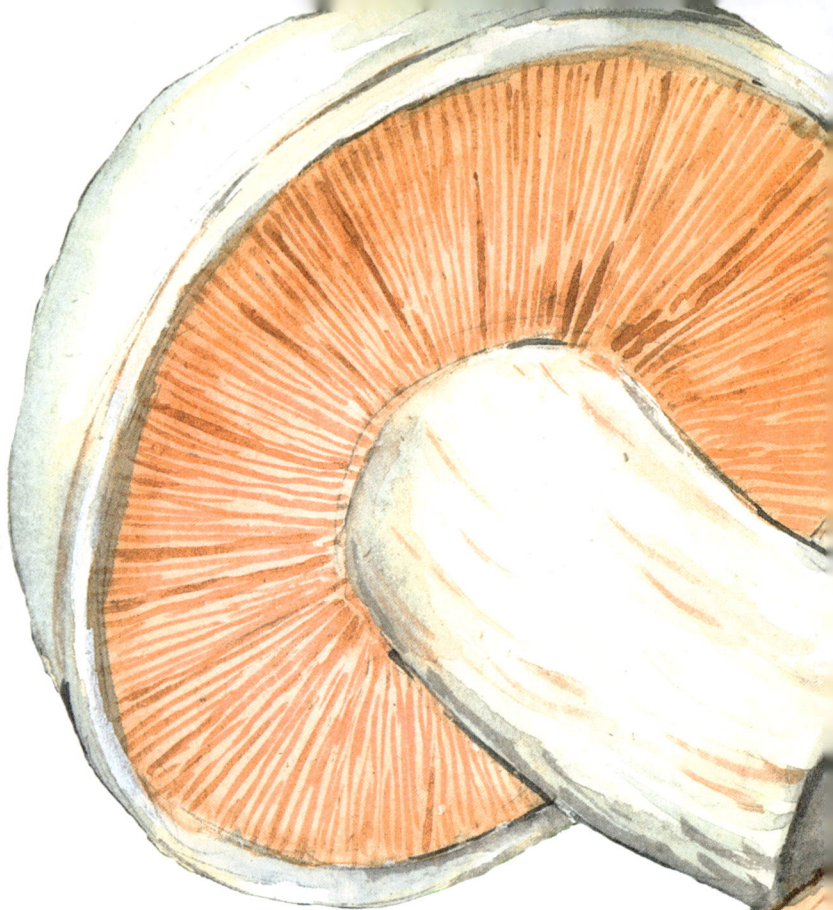

Some have caps with GILLS . . .

Some have caps with PORES . . .

Then there are the

POLYPORES,

which often frill themselves

up tree trunks like a series of shelves.

These can have pores or gills . . .

Others don't have typical
caps at all—

instead, like the **LION'S MANE**
    or magenta coral, they take
the form of a furry, friendly
    monster wondering the day
        away or a deep-sea coral
    into which a fish might dart.

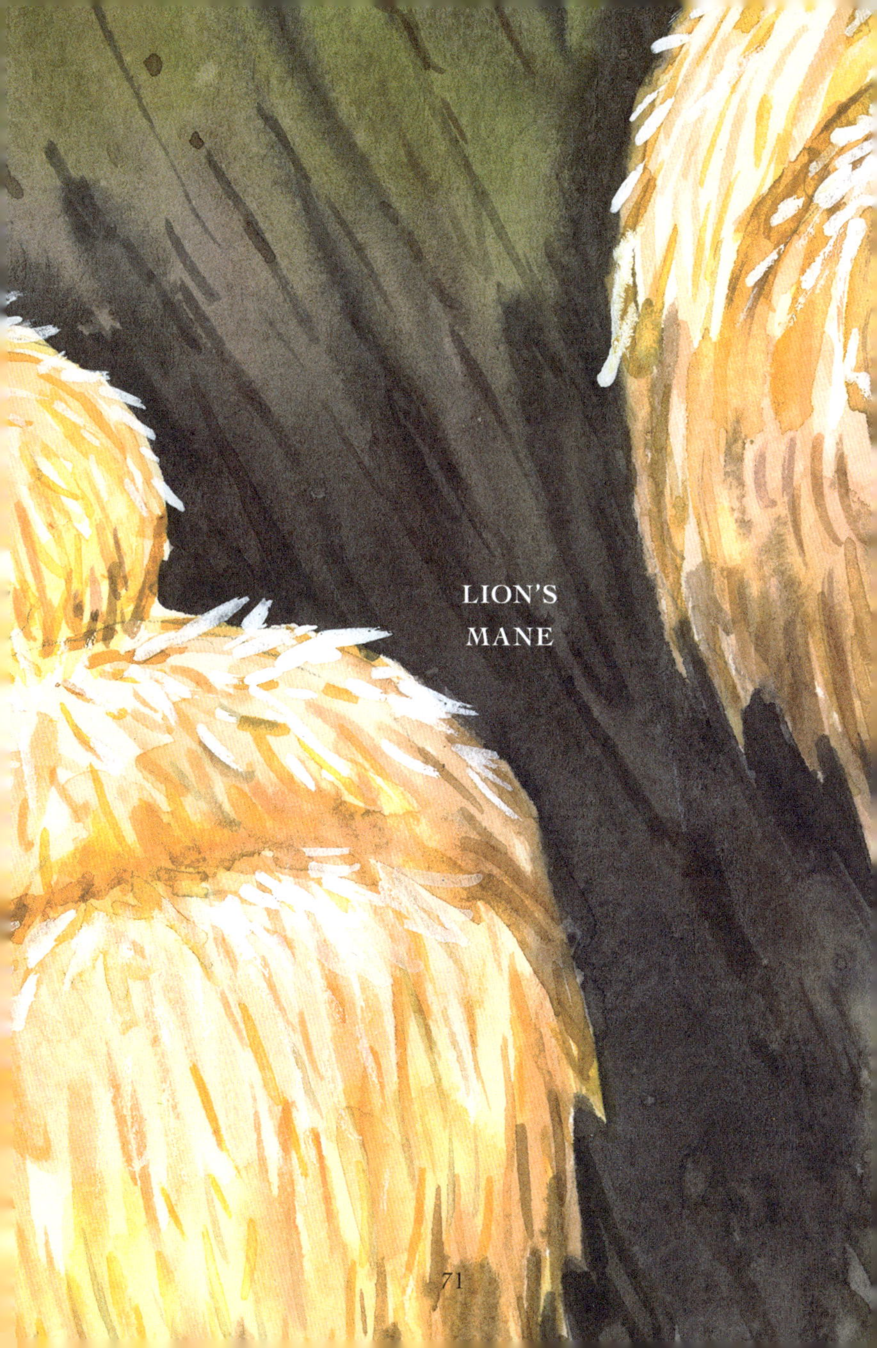

LION'S MANE

After noticing those caps,

    keep noticing—fungi often morph

as the fruit body develops.

Consider those

ROUND, SPHERICAL ones,

    such as the egg phase

of the beautiful (and sometimes deadly)

    Amanitas or puffballs,

famous for emitting a "smoke" of spores . . .

AMANITA (EGG)

PUFFBALLS

or a set of **UMBRELLAS,**

such as mica cap and parasol

CORTINARIUS

PARASOL

the **CONICAL**

purple *Cortinarius*

or inky cap

INKY CAP

75

and, of course,
sometimes FLATTENED,
like someone forgot
to put away the patio furniture,
Amanita, Russula

RUSSULA

a cap like a

FUNNEL,

a ROUND,
THICK-TOPPED bolete,

or tiny, little
PINWHEELS . . .

A thing of beauty

is a joy for ever.

— KEATS

81

OYSTER
MUSHROOM

PORTOBELLO

CRINKLE SKIRTS TO A FAN THAT FRILLS: GILLS

Some of those caps have beautiful gills,
    which hold spores. Others resemble
a crinkle skirt or a fan that frills.
    And these are as diverse as the
caps that wear them.

Notice the GILL STRUCTURES
of these examples and make lists
of details and patterns:

WAVY,

like those in an oyster mushroom . . .

CROWDED,

like those of the portobello mushrooms
from the store . . .

SPLIT
GILL

DIVIDED OR SPLIT, like the split gill . . .

SPREAD OUT/OPEN

like this . . .

PINWHEEL
MUSHROOM

AND SOME HAVE PORES...

Instead of gills, some mushroom
caps contain pores,
which release spores.
The pore shapes take
wild variations . . .

ORANGE PORE
FUNGUS,
an invasive species
now found in Hawaii

ANGLED, looking like skin under a microscope,
such as the lumpy bracket

LABYRINTHIAN, or MAZE-LIKE,

TIGHTLY SPACED, such as the
purplepore bracket with a hairy, whitish cap
and a purple underside.

And then, there are the mysterious ones,
such as the **ARTIST'S CONK,**
with tiny, hard-to-see pores, a favorite of
folk artists who score the surface
to make an etching.

PURPLEPORE BRACKET

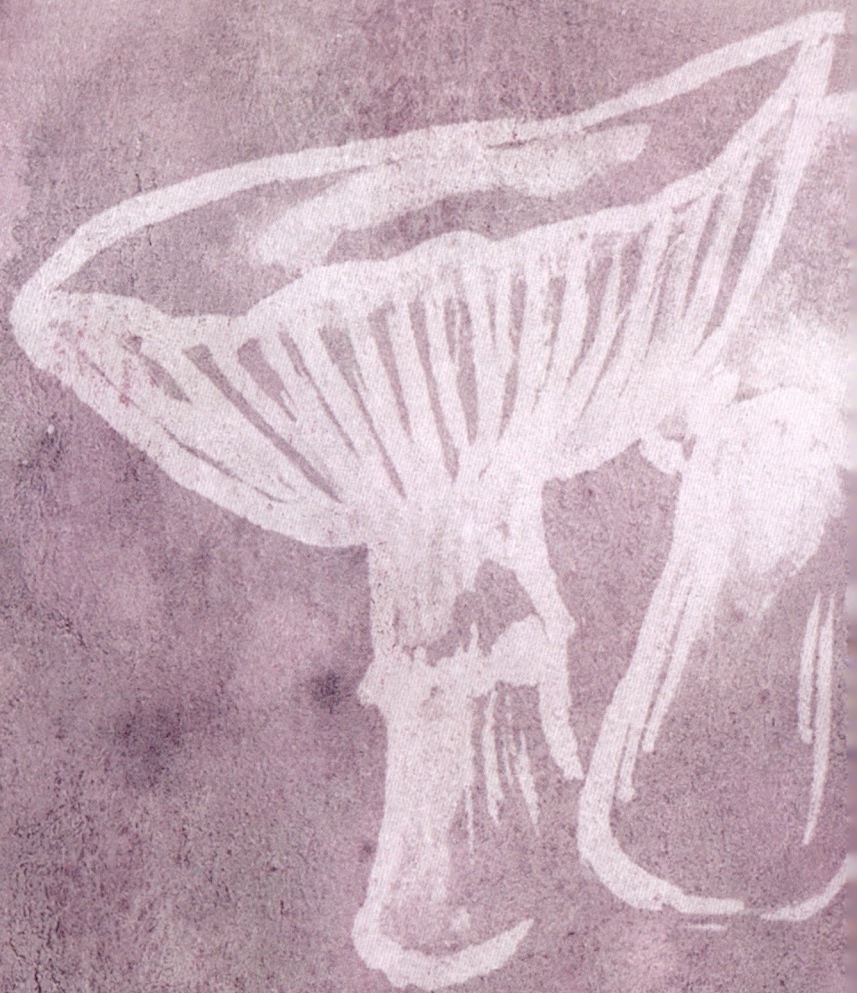

I am waylaid by Beauty.

—Edna St. Vincent Millay

SPORE PRINTS

AS UPSIDE-DOWN FLOWERS

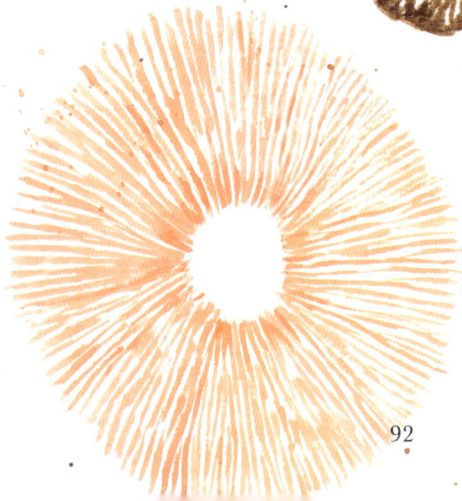

When it comes to fungi,
  ART AND SCIENCE
        often overlap,

whether one is drawing on a conk
    or creating spore prints, which some
    artists like to layer into a collage.

To make a print, simply lop off
    a mushroom's cap, place it gill-
or pore-side down on a piece of paper
        (dark paper for lighter spores,
    light paper for darker spores),
put a bowl or glass lid over it,
    and let it sit overnight.

Remove the bowl in the morning.

The spores will leave a
    gill- or pore-shaped impression
 that mirrors the
mushroom cap's underside,
    like an upside-down flower.

After gathering a few,
   YOU'LL CREATE A BOUQUET
perfect for framing.

And for foragers,

a mushroom's **SPORE COLOR**

is often essential in

helping identify the mushroom.

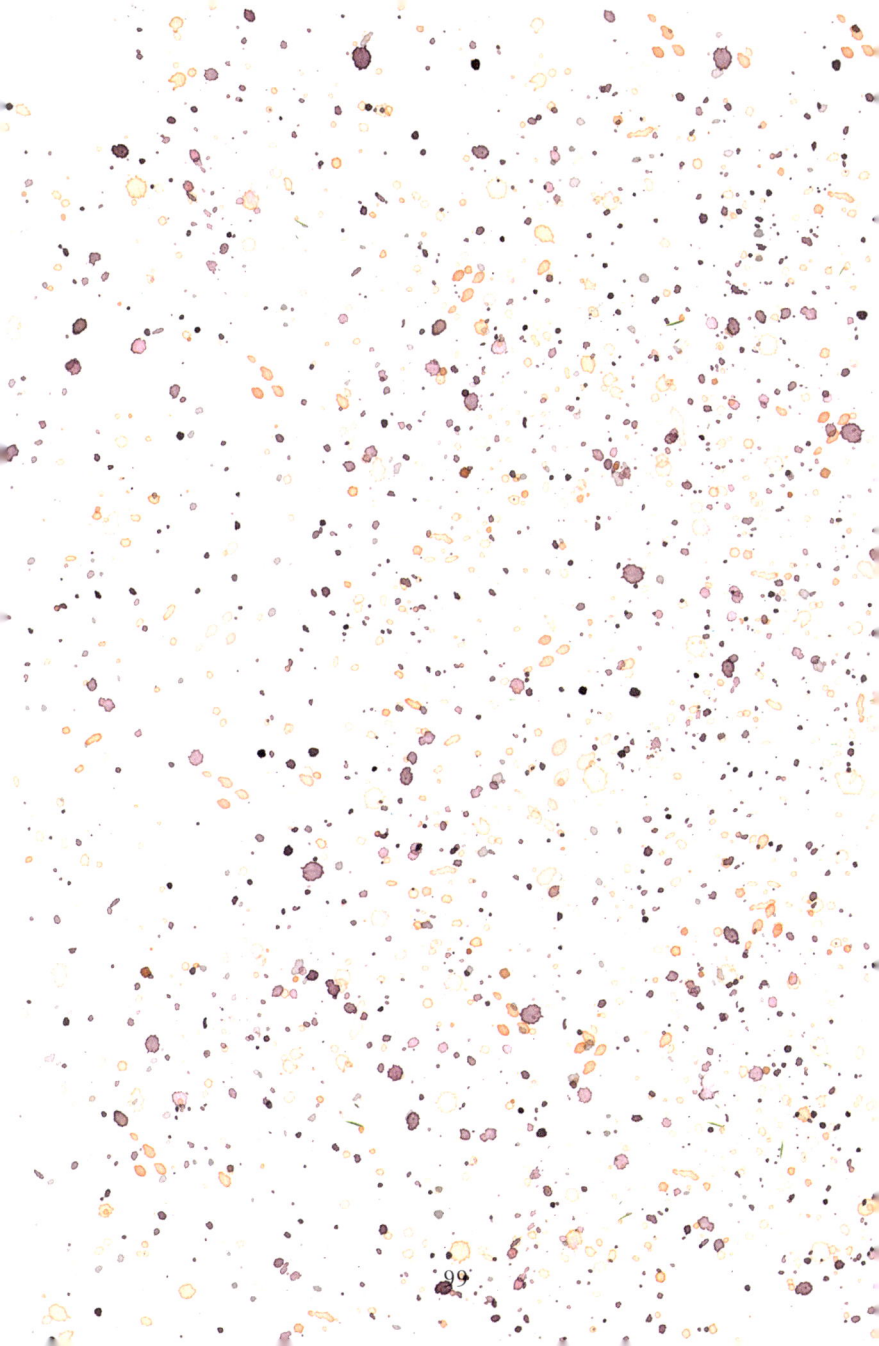

If a shower drives us for shelter to the
maple grove or the trailing branches
of the pine, yet in their recesses with
microscopic eye we discover some new
wonder in the bark, or the leaves, or
the fungi at our feet . . . We can study
Nature's nooks and corners then.

—THOREAU

A WEB BENEATH THE FOREST FLOOR

Most fungi are connected
through MYCELIA,

    filament on filament on filament—
invisible to us above the ground,
        but beneath us nevertheless,
      like an underground cobweb.

These **mycelia** are composed
of threads of HYPHAE:

   tiny, microscopic structures, which,
if you zoom in,
         resemble undersea coral
      in their branches and layers.

Lift a chunk of wood or an old rock,
and reveal the **webby network**
stuck to the bottom, sending out
their little feelers.

This often-unseen **"web"**
is what makes possible the protruding fruitbody
that pops up and delights us:

THE MUSHROOM.

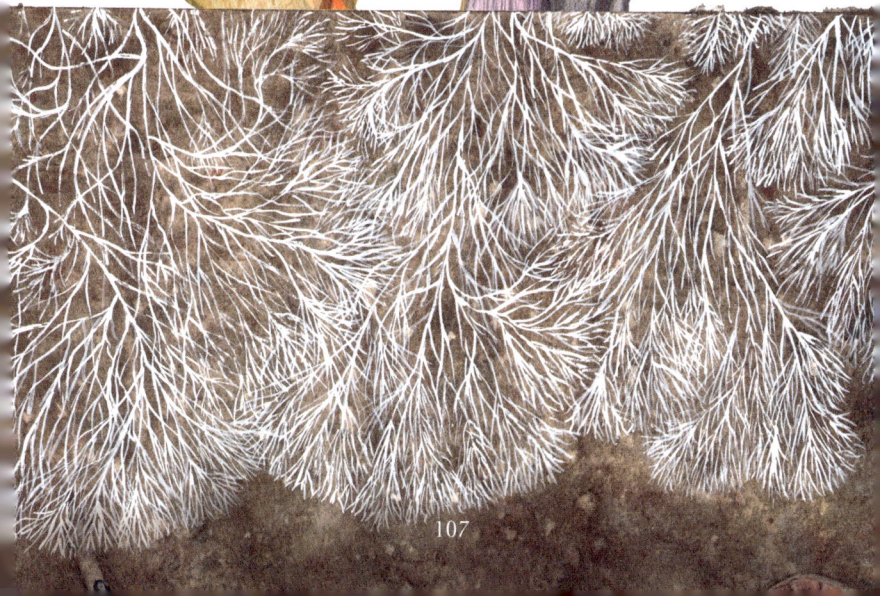

The **life cycle** of a

fruitbody-producing mushroom

is complicated,

but there are three basic phases:

FRUITING,

RELEASING SPORES,

and GERMINATION.

108

A SIMPLIFIED LOOK AT A MUSHROOM'S LIFE CYCLE

Such mushrooms release spores,
which contain all the material
to make a new mushroom and are

**SPREAD BY WIND, INSECTS,
AND OTHER ANIMALS.**

If a spore lands on a choice spot,
hyphae can form mycelium
and create a new environment
for the mushroom to emerge again.

Besides making the whole forest floor possible,
the hyphae that compose mycelia and
mushrooms are also being explored
by engineers and architects.

BETTER THAN PLASTICS

(because totally biodegradable), the glue-like fibers

contain the same chemical structures

that make up the exoskeletons of

crustaceans and arthropods

and could one day help hold

together and insulate houses:

A WAY TOWARD
A GREENER,
SAFER FUTURE.

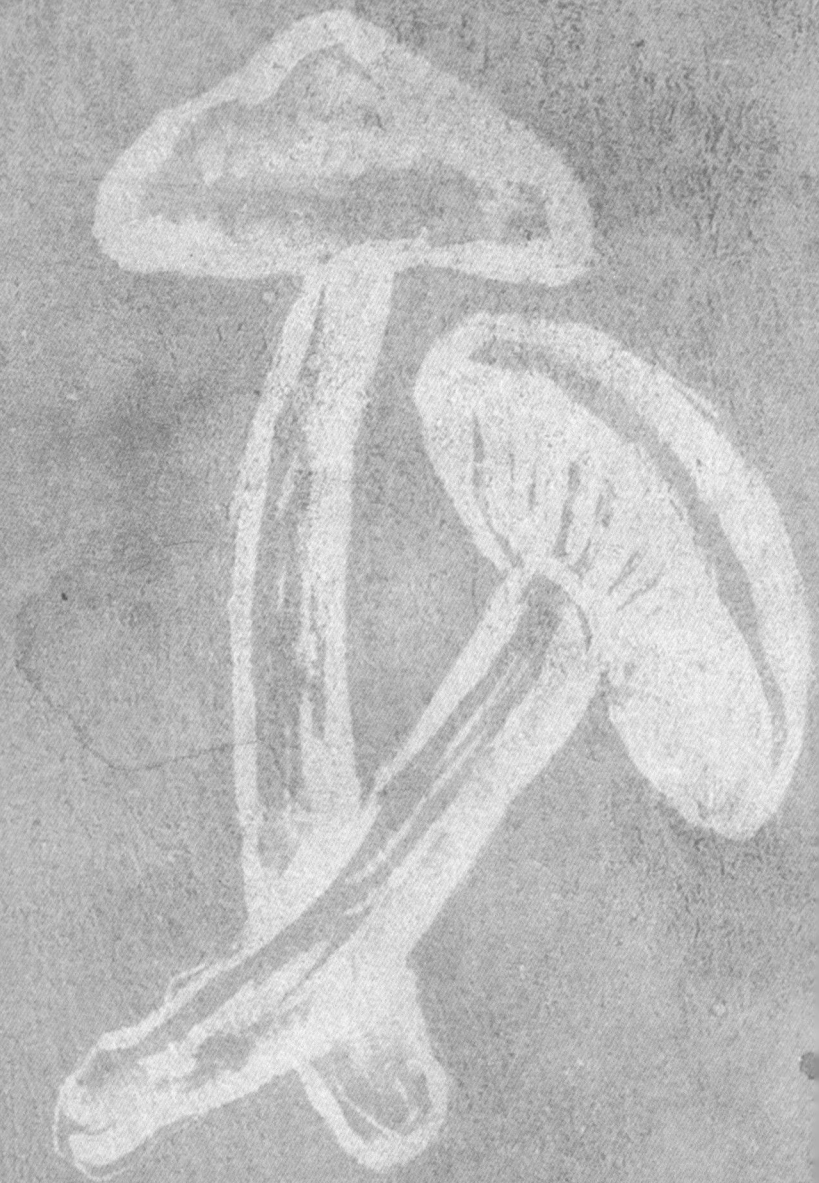

Once we are led away by some
winding pretense of a path, each leafy
curve of which is more enticing than
the last, we are apt to yield ourselves
to the simple charm of being.

—FRANCES THEODORA PARSONS

MYCELIAL MULTITUDES

In parks, forests, even the castoff lots shot through with
"weeds" or a straggly copse of trees, there's a
conversation happening beneath your feet,
an underground hum come up through the dun,
a network of fungi and arboreal tangle playing telephone,
under the surface of the earth.

And the mushroom you see

emerging from leaf litter

LIKE AN ISLAND

jutting from sea is akin to a fruit.

Mushrooms fruit out of the rest
of the underground organism,
which is made up of a whole system
of buried mycelia, threads or root-like fibers
found just below the soil, growing through
the earth and decaying wood,
sending signals to and from tree roots and
other parts of the large mushroom network.

These networks contain

MULTITUDES.

Mycelia send forth mushrooms,
which in turn let loose spores,
small cells that enable reproduction,
allowing the networks to spread
and form in new places.

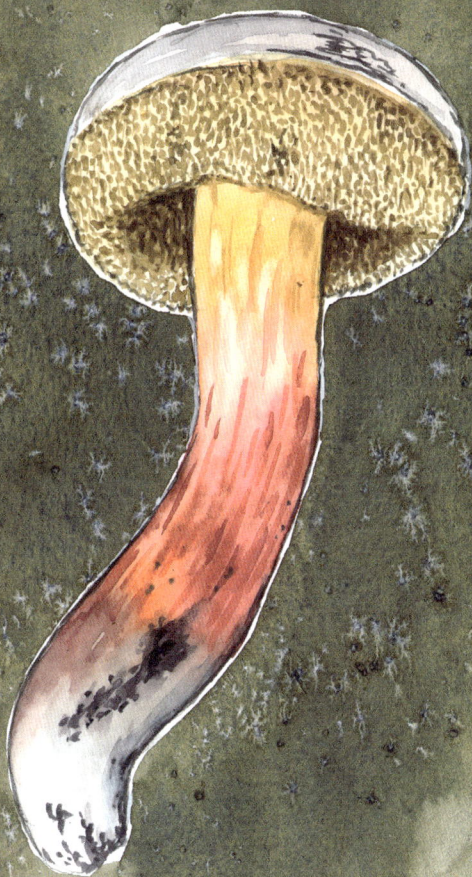

And in some cases, those same underground
mycelia may even help trees and/or
fungi **share resources** such as
water and nutrients.

Just how much trees and fungi "talk"
is a hot-button topic—
and the research is young—
but it's clear enough that
mycelia are their own
understudied, wild places.

All death in nature is birth.

—FICHTE

THE WARM AND NECESSARY BREATH OF DECOMPOSITION

In compost, a host of
organisms, including bacteria
and fungi, make the warm
and necessary breath of
decomposition possible
by breaking down
organic matter.

Mushrooms often have similar jobs,
but writ large:

                      many are **SAPROBES,**
        or **decomposers.**

This unseen, tiny cleanup crew
   breaks down wood and softens trees,
making space for birds
      to build their nests.

They also keep leaf litter
   and other organic debris
in check, turning death into
      fertile ground for new life . . .

Other fungi are **MYCORRHIZAL**
and live in a symbiotic relationship
with plants and trees.

The details are complex,
but the fungal mycorrhizae connect
with plant roots, enabling nutrient transfer,
possibly even serving as
communication networks.

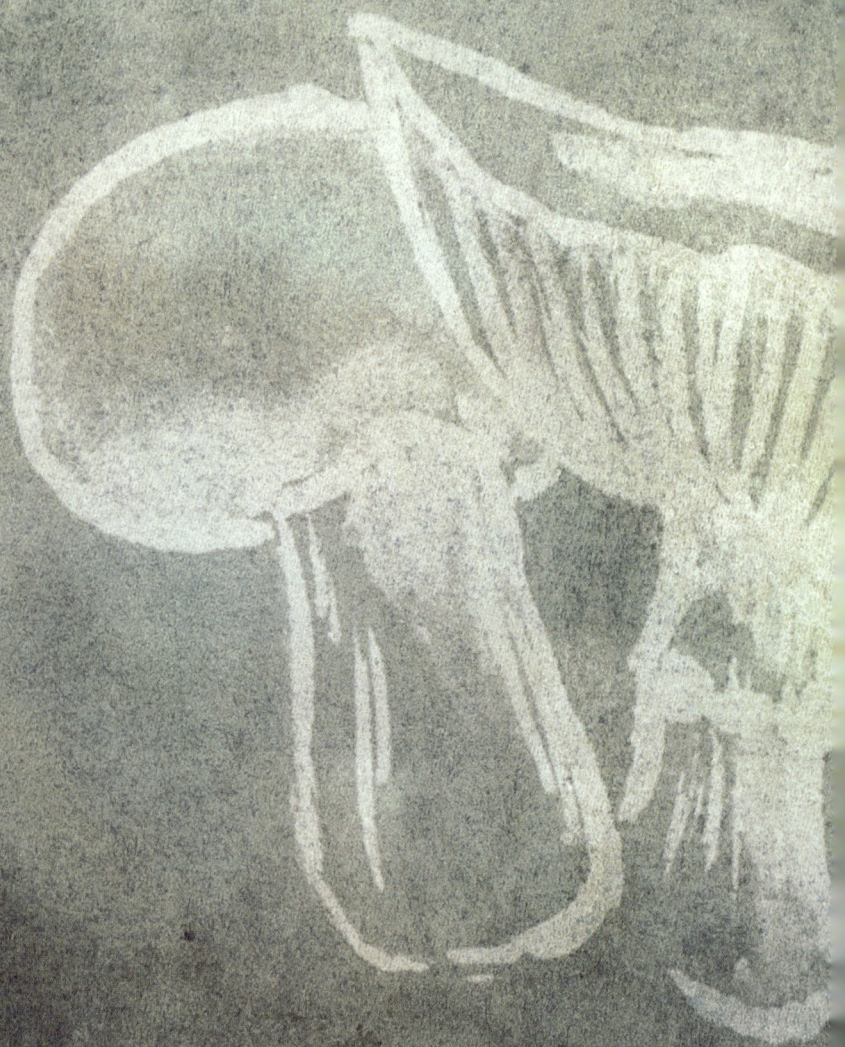

Nature will bear the closest inspection.
She invites us to lay our eye level with
her smallest leaf, and take an insect
view of its plain.

—THOREAU

WORLDS UNTO THEMSELVES

Mushrooms and lichens
are their own worlds—
and they are home to
many types of insects
and other invertebrates.

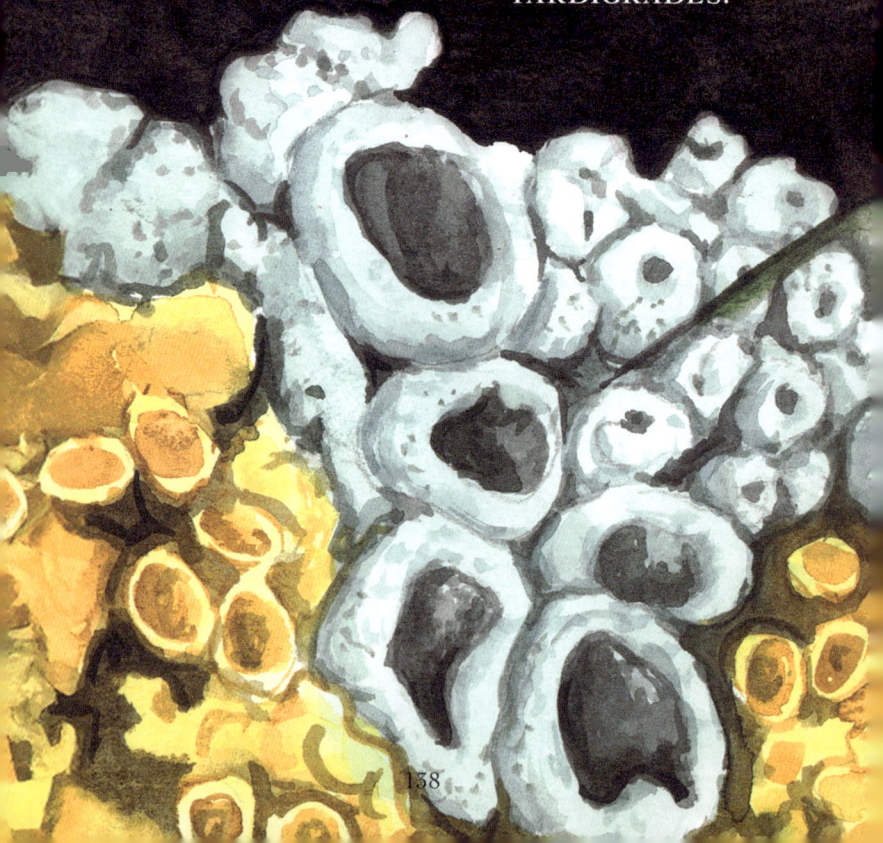

The relationships between fungi
and invertebrates can vary from outright
parasitism to everyone-benefits-mutualisms.
That's just the surface level. If one looks closer
(say, with a microscope) at lichen,
those odd fungi-bearing organisms, they become
environments unto themselves, home to microscopic mites,
nematodes, rotifers, and
TARDIGRADES.

(But to really see the critters
in question, you need to soak
  the lichen in water for some hours

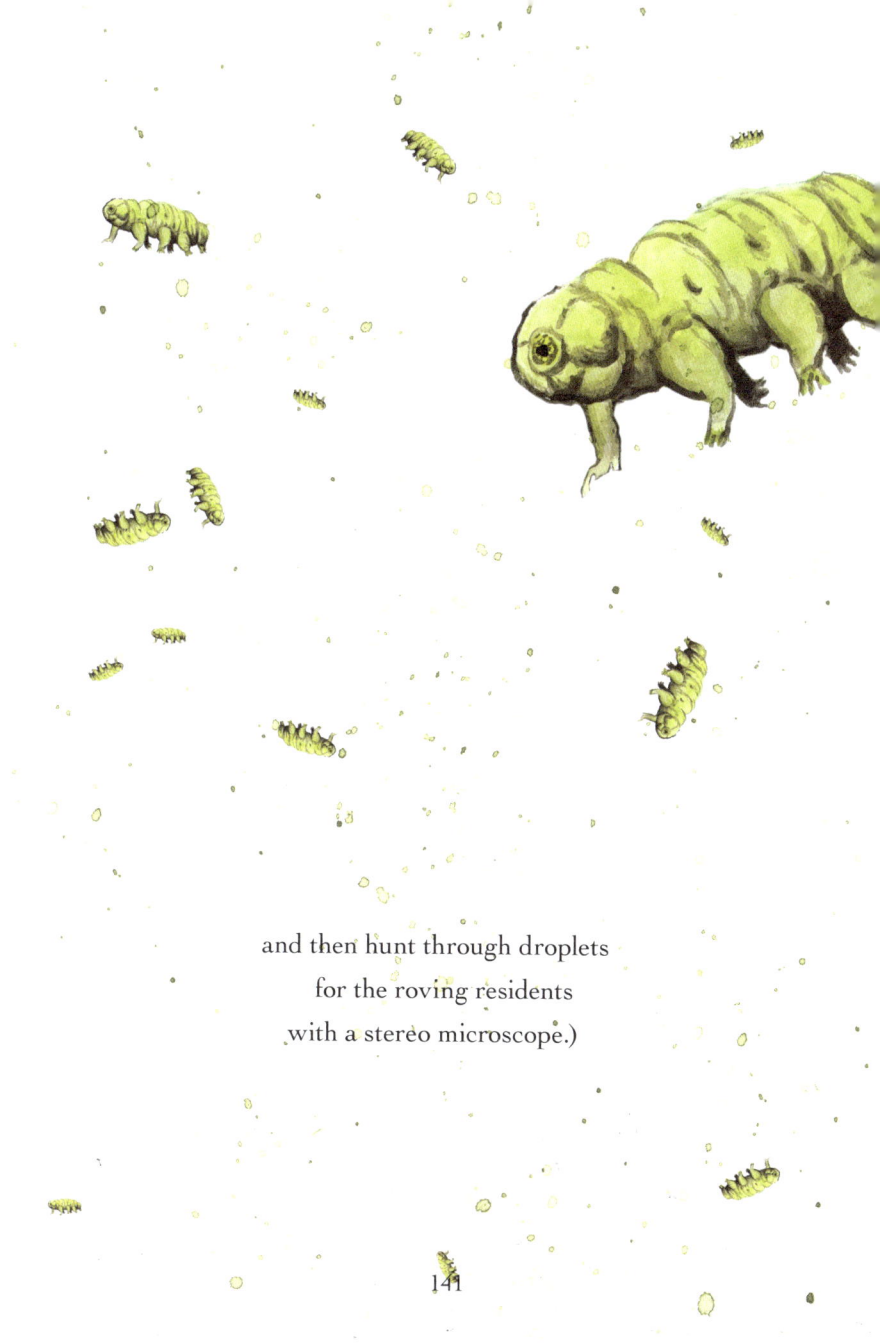

and then hunt through droplets
for the roving residents
with a stereo microscope.)

*I will study the botany of the mosses
and fungi on the decayed [wood], and
remember that decayed wood is not old,
but has just begun to be what it is.*

—Thoreau

A FECUND CELEBRATION OF DEATH

In the Blue Mountains of Eastern Oregon,
      amid about 1.5 million acres of sage desert,
   grasslands, and fir and pine, you'll find the
HUMONGOUS FUNGUS.

It is the largest colony of *Armillaria ostoyae*,
   also known as honey mushroom, in the world.

The five parasitic colonies that
make up this behemoth
**COVER OVER 2,385
ACRES OF FOREST,**
living off the roots of conifers
for thousands of years.

Even though they are so large,
these colonies are mostly found underground,
as mycelia.

But in fall, the honey mushrooms pop up,
covering the forest floor
(and sometimes erupting through
the lower bark of trees) in golden brown.

Honey mushrooms often emerge in clusters,
a fecund celebration of a tree's death:
**via dehydration and starving
it of its own nutrients.**

When the fruitbodies aren't present,
you can still find the mycelium
by peeling back the bark of trees that look
like they're dying: underneath,
a paper-thin, white fungus coats the inner trunk.
Sometimes sap oozes from a tree
as the fir tries to fight off its intruder.

This huge organism has some competition
as the largest on Earth: a stand of
quaking aspen, sometimes called PANDO.

Found in Utah, around 50,000 individual trees
         form this clonal organism of one trembling
    coven, this collective voice of leaf and limb
    that covers 110 acres, likely weighing
              13 million pounds, less in surface area
         but probably heavier than the
                   HUMONGOUS FUNGUS.

Come forth into the light of things,
Let Nature be your teacher.

—WORDSWORTH

A BRIEF HISTORY OF FOXFIRE

Have you seen
its glow?

More than
112 species
of fungi grow
their own sun.

JACK-O'LANTERN

With a soft green light, these
BIOLUMINESCENT FUNGI
glint like wayward stars
amid the leaf litter and rotting logs.

The Jack-O'Lantern, for example,
is common and widespread,
bright orange with a faint green glow,
if you can spot it at all,
and its mushrooms, toxic.

Foxfire probably stems from the
Old French, fols (false),
    but what could be truer than
lost lights scattered through the forest?

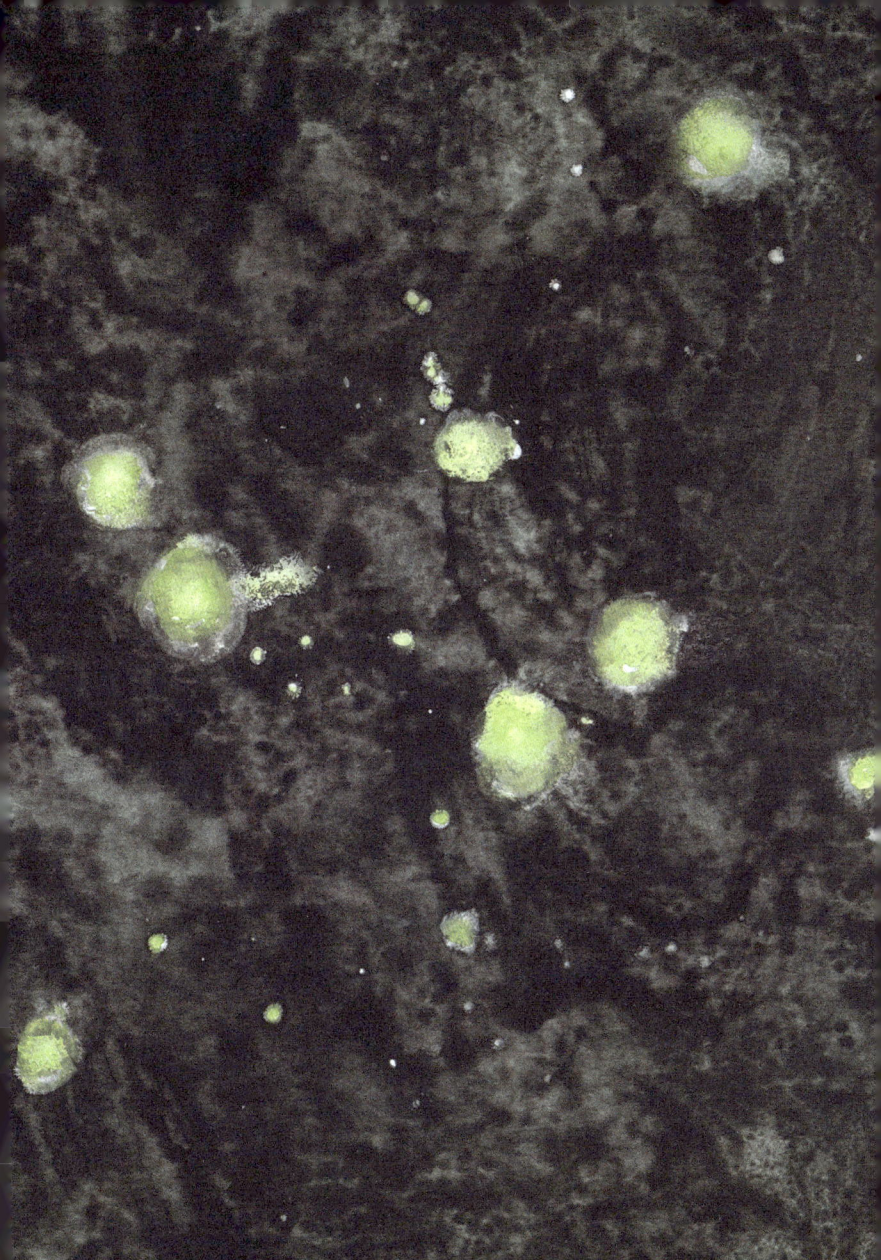

It took until the 20th century before
        scientists finally learned of the cellular
    laboratories that cause foxfire
            to spark to life, but it found observers,
    and uses, much earlier: poets, miners,
        and engineers all remarked upon it,

and well before it was understood,
    crude arrangements were even used to
LIGHT THE DIALS of *The Turtle*,
    the Revolutionary War–era submarine.

(That use, like *The Turtle's* ill-fated
attack run, was a failure.)

Perhaps those particular fungi
      had a say in that after all—
  they are decomposers,
    but not outright destroyers,
        recyclers not raiders.

Who can tell what they speak to the trees,
     the plans the forest and fungi
  make together each night?

Beauty is a good letter of introduction.

—GERMAN PROVERB

TREASURE FROM THE FOREST

The familiar grocery-shelf
mushrooms are a misleading
introduction to the world of fungi.

Button mushrooms, creminis,
and portobellos are all the same
species, just different ages,
youngest to oldest.

But walk a farmer's market with
local finds fresh from the forest
and witness the treasure of
**TRUFFLES, MORELS,
CHANTERELLES,
BLACK TRUMPETS,
PINK OYSTERS, AND
HEN OF THE WOODS.**

The colors and textures and tastes
there make a mockery of
what most consider a mushroom.

MOREL

Mushroom
Mix

Boletes

Chanterelles

Oyster
Mushrooms

OF
MUSHROOMS
AND
MORDANTS

In 1968, New York sculptor and artist
    Miriam C. Rice began experimenting with
mushrooms as a dye source for fabrics,
    beginning with the gilled mushroom sulphur tuft
or clustered woodlover, which grows
                in yellowish tufts or clusters.

173

Rice then played with **puffballs** and **polypores**
(dyer's puffball and dyer's polypore,
which are now renamed for her),
and popularized using MUSHROOMS FOR DYES.

She went on to write books, including
Let's Try Mushrooms for Color (1974),
and inspired artists, scientists,
and craftspeople worldwide.

With the rainbow of fungi available, mushrooms
(and even lichen) are now often sourced for dyes.
The recipe for doing so begins best with older fungi,

and establishing a **BALANCE OF 1:1
FOR FUNGI TO FIBER** (for example,
a pound of violet webcap to a pound of wool).

The mushrooms are dried and then ground up
        (if fresh, they should be chopped
as small as possible to expose more surface area).

    The fibers should be pretreated
  to soak up a **MORDANT**
        (an acid, alum chloride, or mineral
          that helps fabric absorb colors),
  and then soaked
     in a dye bath of mushroom.

WOOL, SILK, COTTON, FLAX, HEMP,
GRASSES, AND HAIR . . .

... all can absorb mushroom hues, which range
from yellows to browns and greens to purples.

Nature alone is antique, and the oldest art a mushroom.

—CARLYLE

WITCHES KNOW WHAT'S WHAT

The Aztecs and Mayans valued visions,
and because mushrooms
produced such seer abilities,
they were highly prized
in pre-colonization cultures.

Mayans even
made artifacts
known as
mushroom stones,

likely a nod to ritual
consumption of
the magic
fruitbodies.

The genus *Psilocybe*, with its

PSYCHOACTIVE COMPOUNDS

known as entheogens,

have been used this way

in Central America

and beyond . . .

with women as the keepers of the

"fungal knowledge," ETHNOBOTANY,

and many spiritual practices.

189

No wonder we turn to midwives and
witches for our magic needs.
Witches know what's what.

The mainstream world
is only now just catching up,
with **MICRODOSING** recently
popular for anxiety, depression,
PTSD, addiction, and other conditions.

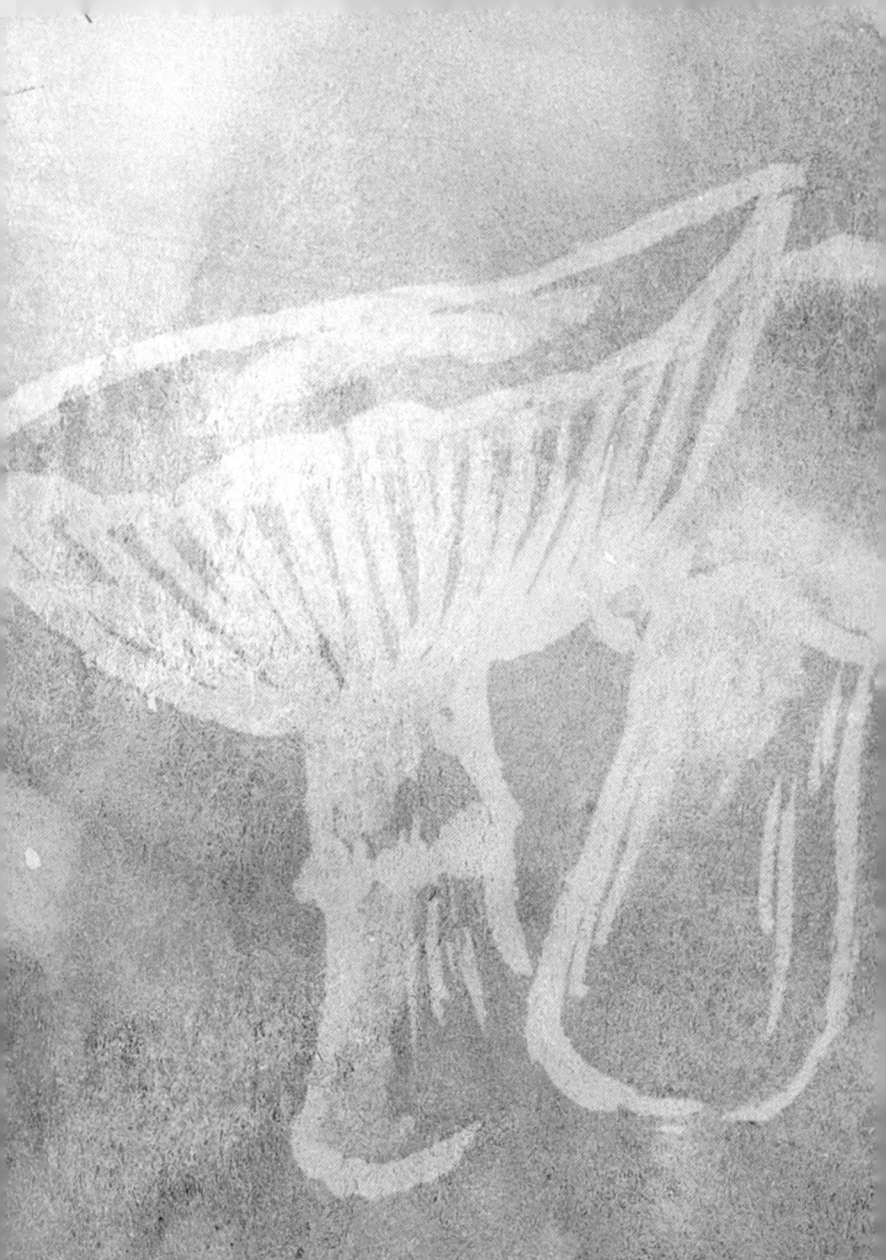

I love delicacy, and for me
Love has the sun's splendor
and beauty.

—SAPPHO

INSIDE THE FAIRY RING

In Shakespeare's *The Tempest*,
Prospero addresses the elves of the hills:

When he comes back; you demi-puppets that
by moonshine do the green sour ringlets make,
whereof the ewe not bites; and you whose pastime
is to make midnight mushrooms . . .

TO MAKE MIDNIGHT MUSHROOMS!

The folklore in Shakespeare's time
was that if you stepped in a fairy ring,
you could die . . .

because the fairies might force you
to dance to death.

In Germany, round marks on grass,
known as **HEXENRINGE,** or witch's rings,
might have been the site of magic spells.

Today, we know such circles form
because of the way mycelium radiates,
pulling nutrients from the soil to
create a ring, releasing nitrogen
as the fungus uses up organic matter
to make the grass around it lusher,
beyond the dead ring . . .
the mycelial front moves forward,
the center decays.

In the Plato-Aristotelian derivation of
SCALA NATURAE, the ladder of being—
which medieval Christians believed to be decreed
by God—mushrooms were near the bottom,
just above minerals (making mushrooms and rocks
the least important and most dirty and suspicious),
as opposed to the divine, or spirit,
which was of the heavens.

However,
some tellings cast
the "lower materials"
as god-like, since
ALCHEMY
promised to turn
them to spirit,
or gold.

All told, fairy ring mushrooms
hold magic and mystery,
as well as history: A group of
TROOPING FUNNEL MUSHROOMS
in Belfort, France, is likely over
700 years old.

And beauty is a flame, where hearts, like moths,
Offer themselves a burning sacrifice.

—OMAR KHAYYĀM

AN UNEXPECTED SAVIOR?

Mushrooms may save us altogether.
Some fungi can **sequester
pollutants,** such as heavy metals.
Others may **break down microplastics,**
and mycoforestry (growing edible fungi
in a forest) may provide
a **carbon-negative food source.**

208

THE NUTRITIONAL
    AND MEDICINAL USES
  OF FUNGI ARE
OLDER THAN HISTORY.

Indigenous peoples,
especially in the Americas,
      have long used fungi in
    sacred rituals,
and cultures from ancient Rome
    and pharaonic Egypt to China
have elevated and revered fungi.

Ask your elders
      what they can share
  of long-cherished uses,
          or more-modern ones.

Imagine our planet rescued
by these **little soldiers of
environmental justice—**
a field of fairy parachutes,

golden spindles, a bluing bolete,
earthballs, earth stars, white coral
fungus arriving as angels to
ferry us all into an age where we

SURVIVE AND THRIVE.

A SAMPLING OF WORDS FOR "MUSHROOM" FROM AROUND THE WORLD

Pilz	German
Champiñón	Spanish
Wazhashkwedowens	Ojibwe
Caŋnakpa	Dakota
Qugam tutusii	*a toadstool,* Aleut
Whareatua	*field mushroom,* Māori
蘑菇	*Mógū, mushroom,* Mandarin
キノコ	*Kinoko, mushroom,* Japanese
Champignon	French
Fungo	Italian
Uyoga	Swahili
Гриб	*(grib),* Russian
Kukuramutta	Hindi
Olu	Yoruba

MUSHROOM NAMES

Many mushrooms have evocative, whimsical common names.
Here is a sampling.

Big Laughing Gym

Black Jelly Roll

The Blusher

Dead Man's Fingers

Destroying Angel

Devil's Snuff Box

Dog Nose Fungus

Dryad's Saddle

Elfin Saddles

Fairy Bonnet, Fairy Cup,
Fairy Fans, Fairy Fingers,
Fairy Helmet, Fairy
Parachutes
(all different mushrooms)

Funeral Bell

Hairy Panus

Jack o' Lantern

Jelly Babies

Jelly Tooth

King Alfred's Cakes

Lawyer's Wig

Leafy Brain

Old Man of the Woods

Pigskin Poison Puffball,
Skull-shaped Puffball

The Sweater

Tree Ear

Trumpet of Death

The Vomiter

Witch's Butter,
Witch's Hat

Wood Ear

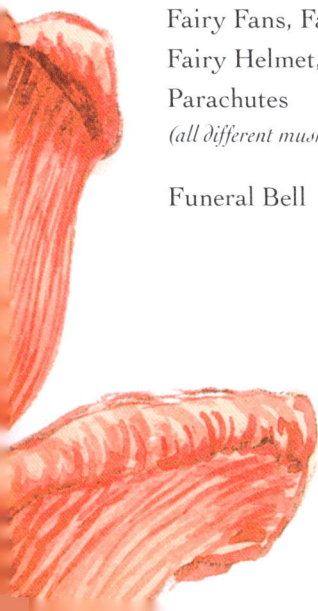

Maya Jewell Zeller is a poet, essayist, educator, editor, and recipient of fellowships and residencies from the Sustainable Arts Foundation, Centrum, Artist Trust, University of Oxford, and the H.J. Andrews Experimental Research Forest. Her most recent poetry collection, *out takes/ glove box*, was selected by Eduardo Corral as winner of the 2022 New American Poetry Prize; she is also the author of the memoir-in-essays *Raised by Ferns* (Porphyry Press); coauthor, with Kathryn Nuernberger, of *Advanced Poetry: A Writer's Guide and Anthology* (Bloomsbury); and coeditor, with Sharma Shields, of *Evergreen: Grim Tales & Verses From the Gloomy Northwest* (Scablands Books). Maya teaches writing and publishing for Central Washington University, as well as poetry and nature writing for Western Colorado University's low-residency MFA program. She lives in the Inland Northwest with her children, with whom she likes to paddleboard, hike, read, play, and get as close to the forest floor as possible.

ABOUT THE ILLUSTRATOR

Jenny deFouw Geuder is an artist and educator from Michigan with bachelor's and master's degrees in art education (and a minor in English). She has taught art at the middle-school level for 20 years and has continued her own artistic interests on the side, both in commissioned work and personal topics. She is the award-winning author and illustrator of *Drawn to Birds: A Naturalist's Sketchbook* and *Nature Explorer*. While she primarily works in watercolors, she also enjoys oils, ceramics, and graphite. She lives in the country with her husband, two boys, a dog, four cats, a hedgehog, chickens, and two horses.

The Story of AdventureKEEN

We are an independent nature and outdoor activity publisher. Our founding dates back more than 40 years, guided then and now by our love of being in the woods and on the water, by our passion for reading and books, and by the sense of wonder and discovery made possible by spending time recreating outdoors in beautiful places.

It is our mission to share that wonder and fun with our readers, especially with those who haven't yet experienced all the physical and mental health benefits that nature and outdoor activity can bring.

#bewellbeoutdoors

WORKS CONSULTED

Arora, David. *All that the Rain Promises, and More: A Hip Pocket Guide to Western Mushrooms*. Ten Speed Press, 2003.

Sept, J. Duane. *Common Mushrooms of the Northwest: Alaska, Western Canada & the Northwestern United States*. Sechelt, BC, Canada: Calypso Publishing, 2012.

FURTHER READING

Marrone, Teresa, and Drew Parker. *Mushrooms of the Northwest: A simple guide to common mushrooms*. Cambridge, MN: Adventure Publications, 2019.

Marrone, Teresa, and Kathy Yerich. *Mushrooms of the Upper Midwest: A simple guide to common mushrooms*. Cambridge, MN: Adventure Publications, 2020.

Marrone, Teresa, and Walt Sturgeon. *Mushrooms of the Northeast: A simple guide to common mushrooms*. Cambridge, MN: Adventure Publications, 2016.